SPACE

'That's one small step for a man,' said the American astronaut Neil Armstrong, and he walked into history – one of the first two men to walk on the Moon. More than forty years later, people still remember this exciting moment.

But our adventures in space have not stopped. Wonderful pictures come to us from millions of kilometres across the universe, and scientists find new planets, new stars, and even new galaxies. We learn more and more about the past, and how the universe began. At the same time, our spacecraft and telescopes travel further and further into space. Will it be in our lifetime that people say, 'I remember when the first astronauts landed on Mars'?

OXFORD BOOKWORMS LIBRARY
Factfiles

Space

Stage 3 (1000 headwords)

Factfiles Series Editor: Christine Lindop

TIM VICARY

Space

OXFORD UNIVERSITY PRESS

OXFORD
UNIVERSITY PRESS

Great Clarendon Street, Oxford, OX2 6DP, United Kingdom

Oxford University Press is a department of the University of Oxford.
It furthers the University's objective of excellence in research, scholarship,
and education by publishing worldwide. Oxford is a registered trade
mark of Oxford University Press in the UK and in certain other countries

ISBN: 978 0 19 423673 7

A complete recording of *Space* is available on CD. Pack ISBN: 978 0 19 423665 2

Printed in China

Word count (main text): 9,311

For more information on the Oxford Bookworms Library,
visit www.oup.com/elt/gradedreaders

ACKNOWLEDGEMENTS

Cover image: Alamy Images (Space/Stocktrek Images, Inc).
The publishers would like to thank the following for their permission to reproduce photographs:
Alamy Images pp.17 (Jupiter and Earth artwork/ Victor Habbick Visions/ Science Photo
Library), 37 (Colonised Mars artwork/ Victor Habbick Visions/ Science Photo Library);
Bridgeman Art Library Ltd pp.14 (The Frost Fair of the winter of 1683-4 on the Thames,
with Old London Bridge in the Distance. c.1685 (oil on canvas), English School, (17th
century)/ Yale Center for British Art, Paul Mellon Collection, USA), 38 (Galileo Galilei (1564-
1642) 1858 (oil on canvas), Keler-Viliandi, Ivan Petrovich (1826-99)/ Regional M. Vrubel
Art Museum, Omsk); Corbis p.15 (Sun setting over rice field/ Yasuko Aoki/ amanaimages);
NASA pp.0, 6, 47, 56; Science Photo Library pp.2 (VLA radio antennas/ Tony Craddock),
5 (Orion and Sirius over Iran/ Babak Tafreshi, Twan), 8 (Orion nebula/ Robert Gendler),
9 (Supernova explosion/ Leonhard Scheck), 11 (Solar prominence/ SOHO/ ESA/ NASA),
13 (Solar chromosphere/ Greg Piepol), 18 (Solar system orbits/ Detlev Van Ravenswaay),
20 (Mercury/ NASA/ JHU-APL/ ASU/ Carnegie Institution of Washington), 21 (Venera 4 Soviet
space probe/ RIA Novosti), 22 (Venus surface/ NASA), 23 (Northern lights/ Jeremy Walker),
24 (Mount St. Helens ash plume/ David Weintraub), 25 (Orbits of the Earth and Moon/ Gary
Hincks), 26 (Artwork of theory of Moon's origin/ Joe Tucciarone), 28 (Edwin Aldrin walking on
moon/ NASA), 30 (Eugene Cernan on Lunar Rover/ NASA), 32 (Schiaparelli's observations of
Mars/ Detlev Van Ravenswaay), 34 (Mars exploration rover/ NASA), 35 (Martian volcano/ NASA),
38 (Galileo telescope/ Gianni Tortoli), 39 (Jupiter/ NASA), 41 (Saturn/ NASA/ ESA/ STSCI/ R.
Beebe, New Mexico State U.), 43 (Voyager 1 launch/ NASA), 45 (Neptune/ NASA), 48 (Halley's
comet/ Harvard College Observatory), 49 (Europa/ NASA), 50 (Europa's surface/ NASA),
52 (Titan/ NASA), 54 (Kepler Mission primary mirror inspection/ NASA/ Ball Aerospace),
55 (Diagram of Voyager 2 route/ NASA).

CONTENTS

Earthrise

The three men were alone. Nobody could see them, nobody could talk to them. No one had ever been here before.

It was 24 December 1968. American astronauts Frank Borman, James Lovell and William Anders were behind the Moon in a spacecraft called *Apollo 8*.

The sky above them was black and full of stars, millions and millions of stars – too many to count. But the men were not watching the stars. They were watching the Moon as it moved below them.

Everything that they saw on the Moon was grey and dusty and dead. There were no rivers or seas, no lights or colours – no life at all. Just hills and rocks, dust and stones. But they were studying these hills and rocks carefully, because no one had ever seen them before. Then they saw something new.

As they watched, the Earth came out from behind the Moon like the Sun rising in the morning. But it was more beautiful than the Sun. They saw the blue of the seas and the white of the clouds. As they watched, the Earth rose into the black sky in front of them.

No one had ever seen the Earth like this before. Frank Borman took a photo, but it was in black and white. William Anders quickly put a colour film into the camera and took a colour photograph.

This photo, 'Earthrise', is one of the most famous photos ever made. It shows the Earth as a beautiful blue and white ball against the blackness of space. It is not a big ball; it is really quite small. To the three men in *Apollo 8*, the Earth looked no bigger than a man's hand.

But it is our home. The Earth is the planet that we all live on. We cannot live in any other place.

The observatory at Socorro, New Mexico

2 In the beginning

How many stars can you see in the night sky? In a city, you probably cannot see very many. But on a clear night, far from city lights, you can see hundreds, perhaps several thousand. With a telescope, you can see many thousands. With a really good telescope, scientists can see hundreds of millions.

But look closer: as well as stars, there are galaxies out there. What is a galaxy? It is a group of stars – millions, billions, or even hundreds of billions, all in the same part of space.

If you try to count all these stars and galaxies, you reach an impossible number. The scientist Carl Sagan said, 'The total number of stars in the universe is larger than all the grains of sand on all the beaches of the planet Earth.'

So where did all these stars and galaxies come from? How were they made? Today, space is very big – enormously big – but scientists say that between 12 and 14 billion years ago, space was really small, impossibly small. Everything in the universe was pushed together in a space smaller than a full stop.

Then there was a BIG BANG and everything in the universe began to move away from the explosion at enormous speed. (This idea about the beginning of the universe is often called the Big Bang.) There were clouds of hot gas – hydrogen and helium. As they flew through space, over millions of years, these gases began to move around in spirals, like water going out of a bath. Many pictures of galaxies have this spiral shape.

Inside the big spirals of each galaxy there were millions of smaller spirals, which formed into huge burning balls, and each burning ball was a star. And all these galaxies are still moving away from each other at enormous speed, so space is getting bigger and bigger all the time.

So how big is space? Well, it is much bigger than we can easily understand. But, amazingly, we can use our eyes to see how big it is.

Look at the group of stars called Orion. On a clear night, you can see these easily without a telescope. The star at the top left is a red star called Betelgeuse. The star at the bottom right is a blue star called Rigel. Both of these stars are much bigger and hotter than our own Sun. But they look small because they are a very long way away.

The light from the red star – Betelgeuse – has taken 643 years to travel to your eye. And the light from the blue star – Rigel – has taken even longer: 773 years. Light travels at about 300,000 kilometres every *second*, which is 9,460,730 million kilometres a year (we call that distance a light-year). So that blue star, Rigel, is 7,313,144 billion kilometres away. But you can still see it – with your own eyes, without a telescope. That is amazing.

Of course, you cannot see what Rigel looks like today – nobody can. You can only see what the star looked like 773 years ago. So when you look at a star like Rigel or Betelgeuse, you are not just looking at something an enormous distance away; you are also looking back in time.

With a really good telescope, you can see much further, in space and in time. The Hubble Space Telescope has taken pictures of galaxies that are nearly 13 billion light years away. Light left these stars soon after the Big Bang, so we are seeing back in time to the beginning of the universe.

Space is really very, very big.

Betelgeuse

Rigel

Orion

A picture of galaxies from
the Hubble Space Telescope

3   All about stars

What are stars made of? They are enormous balls of hydrogen and helium, and they are extraordinarily hot. But not all stars are the same. If you look closely, some are quite red, like Betelgeuse, and others are blue, like Rigel. Blue stars are the hottest, red stars are a little bit cooler.

Stars do not live for ever. New stars are made all the time, and old stars die, in the most amazing ways.

In every galaxy, there are large clouds of gas, called nebulae. Inside these nebulae, new stars are born. There are small stars, middle-sized stars, and very large stars. Some of these large stars are a hundred times bigger than our Sun. These big stars do not live very long – only about ten million years – but their short lives are very exciting.

Inside a very large star the temperature gets hotter and hotter. Hydrogen turns to helium, and then into other things – carbon, oxygen, and iron. But then, one day, the star becomes too hot, and there is not enough hydrogen left. Then, very suddenly, the star collapses. As it collapses, it gets even hotter still, and then it explodes. In the explosion, the temperature may reach 100 billion °C.

This exploding star is called a supernova. One day in the next million years, Betelgeuse, that big red star in Orion, will explode like this. In fact, it is possible that it exploded a hundred years ago, and is already a supernova. But the light from the explosion has not reached us yet, so we cannot see it.

When the light from a really big supernova does reach the Earth, it is really exciting. Sometimes people can see

A nebula

the supernova in daylight. People in Japan and China saw a supernova like this in 1024 AD, and people in England saw one in 1604 AD.

In the heat of the explosion even more things are made. For just fifteen seconds, the temperature can be hot enough to make gold and silver. As the star gets cooler, a new cloud of rocks and dust and gas appears, making beautiful shapes in the sky.

The dust and gas from exploding stars flies out into the universe. Over billions of years, some of it becomes stars like our Sun. Some of it becomes planets like the Earth. And scientists believe that some of it becomes rocks, rivers, trees, animals, and people.

They believe that everything on the Earth was once part of a supernova. Everything you see, everyone you have ever met, every part of your body, was once part of an exploding star.

We are all made of stardust!

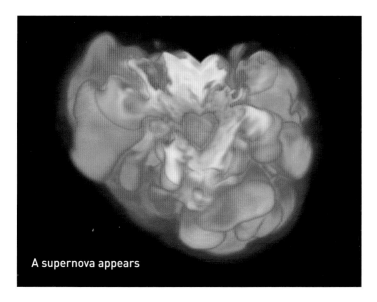

A supernova appears

4 The Sun

Our nearest star is the Sun. The Sun is a middle-sized star. It is bigger than some stars, but much smaller than others. Some stars, like Betelgeuse, are more than a thousand times bigger than our Sun. But we are lucky that our Sun is not too big, because it is not going to explode into a supernova. Not now; probably never. If it does, it will not be for a long, long time.

Middle-sized stars, like our Sun, live for about 12 billion years. The Sun is about 4.6 billion years old now, so it will live for several billion more years. It will go on shining for a long time into the future.

For us, the Sun is the most important star in the universe. Nothing on the Earth can live without heat and light from the Sun. We all know this – people have looked at the Sun for millions of years. Thousands of years ago, scientists in China and Egypt studied the Sun, as they tried to understand why some summers were warmer than others.

All these people asked questions about the Sun. Here are a few of them. How far away is the Sun? How hot is it? And where did it come from?

Scientists today know the answer to the first two questions. The sun is 149 million kilometres away, and it is very VERY hot indeed. No astronaut or spacecraft can get anywhere near the Sun, because it is just too hot.

The hottest fire that we usually use on the Earth is about 2,000 °C. At a temperature like that, iron melts – that is how we make cars and ships and bridges. But that is much

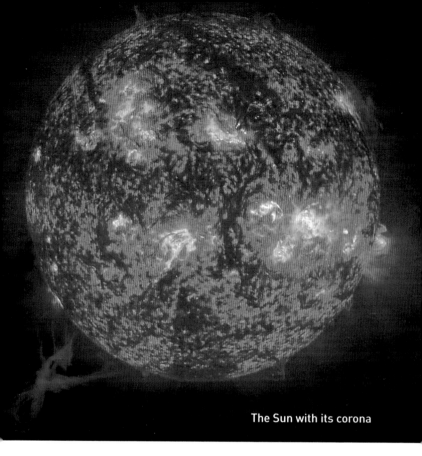

The Sun with its corona

less hot than the temperature of the Sun. The surface of the Sun is about 6,000 °C. But in the Sun's corona – the huge flames just above the surface – the temperature can be 2,000,000 °C. And in the middle of the Sun the temperature is about 15,000,000 °C! That is REALLY hot.

Why is the Sun so hot? What is happening inside it? Well, the Sun is made of just two things, really – hydrogen and helium. All the time, billions of times every second, tonnes and tonnes of hydrogen are turning into helium. Each time they do this, it is like a nuclear explosion. It is like a trillion nuclear bombs going off, every second. At the centre of the Sun, in fact, is an enormous, endless nuclear explosion.

But nuclear bombs are dangerous – they can destroy all of a city. So isn't the Sun dangerous too? Well yes, it is –

but luckily for us, it is also a very long way away. It is 149 million kilometres away – so its heat keeps us warm but does not burn us up. Light from the Sun, travelling at 300,000 kilometres a second, takes just over eight minutes to reach the Earth.

But the Sun does some very frightening things. Never look at the Sun through a telescope – the sunlight will burn your eyes. When scientists look at the Sun, they let the light go through a telescope onto a white wall, or into a computer.

The Sun does not explode, because it is so big – gravity holds it together. But there are enormous flames, called solar flares, above the surface of the Sun. Some of them are 15,000 kilometres high. These solar flares are made by storms on the surface of the Sun, and some of them are as big as the Earth.

Sometimes when there is a really big storm on the Sun, it can start terrible trouble here on the Earth. On 13 March 1989, for example, a cloud of hot gas a million kilometres long came from the Sun towards the Earth. When it came close to the Earth, all the lights went out near Quebec in Canada. A million homes lost electricity, and some were without electricity for eight days, in the middle of winter. Further north, on the same day, people saw huge red and green lights in the sky, and hundreds of satellites stopped working.

Luckily for us, this does not happen very often. Most of the time, the Earth's atmosphere protects us from this danger. But the Sun can change things on the Earth in other ways too. Thousands of years ago, when scientists in China were studying the Sun, they noticed some strange dark spots on the surface. Today we call them 'sunspots'. These sunspots are a little cooler than the rest of the Sun – 4,000 °C instead of 6,000 °C.

Today, these sunspots come and go about every eleven years. But they seem to be important. Between 1645 and 1715 there were almost no sunspots, and during this time the Earth got really cold. In London, people held parties and even lit fires on the frozen River Thames. England in winter was as cold as Russia and Sweden. This was part of a time of colder temperatures called the Little Ice Age, which went from 1550 to 1850.

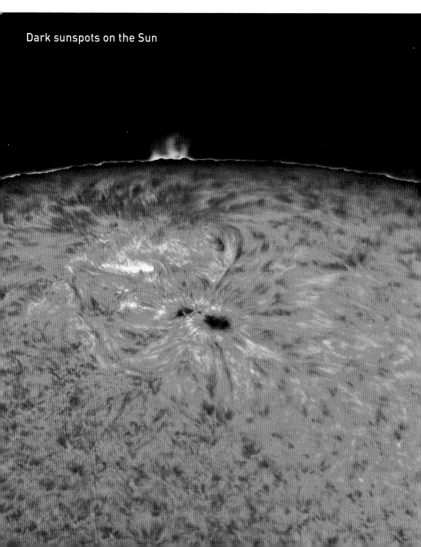

Dark sunspots on the Sun

The frozen Thames
in 1683

And there were other times, millions of years ago, when most of the Earth was covered by ice, many kilometres thick. These were the real Ice Ages. Scientists think that perhaps the sunspots had disappeared then too, and the Sun was cooler than it is today.

So the Sun is enormously important to us. For us, it is the most important star in the universe. But where did the Sun come from? How did it begin? That is the hardest question. And it is like another difficult question: how did the Earth, and all the planets, begin? Where did they all come from? It has taken scientists a long time to find the answers to these questions.

For life on Earth, we need the Sun

5   All about planets

Our Sun, like all the other stars, was formed from a cloud of gas and dust floating in space. Slowly, over millions of years, gravity made the gas and dust come closer together. As it came together, it began to spin in a flat spiral shape, like water going out of a bath.

A round ball formed at the centre of the spiral. The ball was spinning too, and just like water going out of a bath, it spun faster near the centre. As the ball spun faster and faster, it grew bigger and bigger, and hotter and hotter. When it was really hot it started to shine, and this spinning ball of fire became a new star – the Sun.

This was the beginning of our solar system – the Sun, and the eight planets that move around it. Almost all the gas and dust fell into the Sun. Scientists say that 99.8 per cent of everything in the solar system is in the Sun. The Sun is enormous. It is much, much bigger than all its planets added together.

Think of a ball, about 230 mm across – a little bigger than a football. If that ball is the Sun, how big is the Earth, and how big is Jupiter, the biggest planet?

Jupiter is a little ball – just 24 mm across. And you have to look hard to see the Earth – it is a tiny ball, only 2 mm across. That's all. That's how small the Earth is.

So the planets are very small, much smaller than the Sun. The planets are made of just 0.2 per cent of all the dust and gas that went into the solar system. But the planets are very important, of course, because one of the planets – the Earth

Big and small - Jupiter (above) and Earth (below)

– is our home. We all live on one of these tiny planets.

How were the planets formed? They were formed like the Sun, in the same spiral of gas and stardust. While most of the gas and stardust fell into the Sun, 0.2 per cent of it stayed outside. Gravity pulled the rocks together, so that they crashed into each other, and made bigger rocks. The same thing happened to the balls of gas.

You can see this happening to bubbles – balls of air – in bath water. The bigger bubbles pull the smaller bubbles towards them. When the two bubbles touch each other, they join up, and the big bubble gets bigger.

There are eight planets going around the Sun, but they are not all the same. The first four planets – Mercury, Venus, the Earth, and Mars – are all made of rocks. They are quite small, and move quite quickly around the Sun, because they are close to it – like the water near the centre of the spiral in a bath.

Between Mars and the next planet, Jupiter, there are lots of small rocks, called asteroids. Some of them are as big as France, but many are much smaller – the size of a ship or a bus. The asteroids are rocks which did not form into planets.

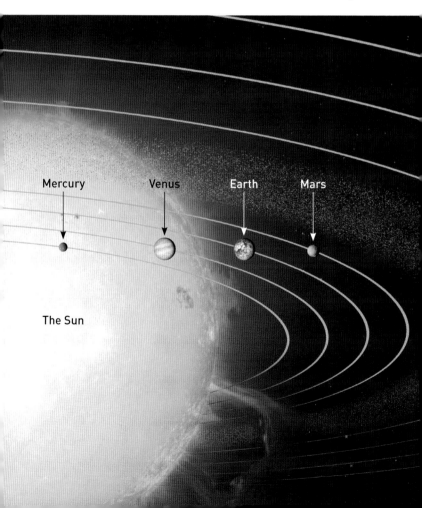

Then there are four more planets – Jupiter, Saturn, Uranus, and Neptune. These are all much bigger; for example, the distance around the middle of Jupiter is eleven times the distance around the middle of the Earth. They are much further away from the Sun, so they take longer to move around it. Jupiter and Saturn are mostly made of gas. Uranus and Neptune are mostly made of ice.

All of these planets and asteroids go around the Sun in the same way. We know from this that they formed as part of the spiral when the Sun became a star.

The solar system

6   Mercury and Venus

Mercury is the planet nearest to the Sun. It is a small planet about the same size as our Moon, and because it is so near the Sun it moves very fast – at about 48 kilometres per second. This means that a year on Mercury – the time it takes to go around the sun – is eighty-eight days.

In 1974 the spacecraft *Mariner 10* flew close to Mercury and took lots of photos. These photos show that the surface of Mercury is covered by craters, just like the surface of the Moon. These craters are places where something – a large rock or a comet – has crashed into Mercury. Some of these craters are huge. One crater called the Caloris Basin is bigger than the five Great Lakes between Canada and the United States! The rock which made this crater made a big hole in one side of the planet and a huge mountain on the opposite side.

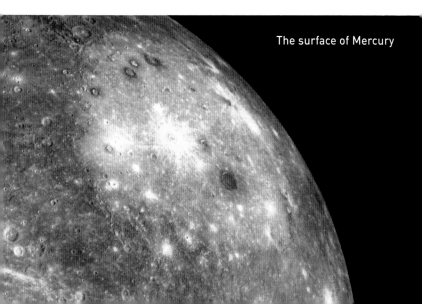

The surface of Mercury

The *Venera 4* space probe

Mercury is very hot because it is so near the Sun. It also has no atmosphere, so it is not a place where humans could live. But for a long time, scientists thought that Venus, the next planet, was a wonderful place. It is the same size as the Earth, and it has an atmosphere too. In 1726 an Italian astronomer, Francesco Bianchini, said that he could see oceans and land there. Scientists thought that strange people lived there – people like us, but bigger, perhaps more beautiful, people with wings!

Sadly, this is not true. In 1967, the Russians sent a space probe, *Venera 4*, to land on Venus. When *Venera 4* floated down through the atmosphere of Venus, the scientists had an unpleasant surprise. Venus was not a nice place at all. The atmosphere was amazingly hot – over 400 °C. A few minutes later *Venera 4* stopped sending messages; it had begun to melt in the heat.

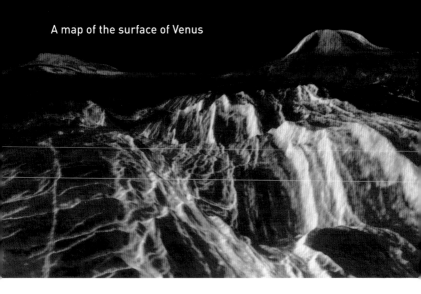

The Russians sent stronger space probes, but the information about Venus just got worse. The air on Venus has no oxygen; it is 95 per cent carbon dioxide (CO_2). The pressure from this atmosphere is ninety times heavier than that on the Earth; it is like the pressure at the bottom of the Earth's oceans. On the ground there are 100,000 volcanoes. There are lakes of molten rock, and molten metal falls like snow onto the surface of the planet. Perhaps once, a long time ago, there was water on Venus, but it has all gone. No humans can possibly live here.

But why is Venus so hot? Scientists think it is because of the carbon dioxide in the atmosphere, and the clouds in the sky. These things keep the planet warm. The heat of the Sun comes in, but it cannot go out. And so the planet gets hotter and hotter.

People on the Earth are worried that this could happen here. Today, about 0.035 per cent of our atmosphere is carbon dioxide. In 1955 it was about 0.0315, so we are adding more carbon dioxide all the time. If we add much more carbon dioxide, scientists say, the Earth will get warmer, and life on the Earth will be more and more difficult.

7   The Earth and the Moon

The most beautiful planet in the solar system is the Earth.

Why is it beautiful? Because it is our home; it is the only planet where humans can easily live. But not just humans; so far as we know, the Earth is the only planet with living plants and animals. Perhaps there are others, somewhere in the universe, but we do not know where they are. No one has ever seen them, although astronomers are always looking around other stars.

Why is there life on the Earth? Because the Earth has an atmosphere with plenty of oxygen, and seas, lakes, and rivers with plenty of water. Both of these things are necessary for life.

The Earth's atmosphere is not just good to breathe; it also protects us from the Sun's most dangerous rays. In the winter, in the Northern Lights, we can see this. The sky turns green and blue as the Sun's rays hit it, and – most of the time – the dangerous rays from the Sun are kept away.

The Northern Lights

Mount St Helens erupts in 1980

The Earth's temperature, too, makes things easy for life. The Earth is not too hot, like Mercury and Venus; and it is not too cold, like Mars and the other planets further away from the Sun. The surface of the Earth is a good place for plants and animals to live, and even more animals live in the sea.

But under the surface, the Earth is very different. Even a small distance underground, the temperature is much warmer. A kilometre underground, where men dig for gold, the temperature can be 40 °C. Further down, where it is even hotter, the rocks begin to melt. When a volcano erupts, we see this red-hot, molten rock come out. Sometimes it moves faster than a car. In May 1980 Mount St Helens, in the United States, erupted with an explosion as big as 1,600 nuclear bombs.

Around the centre of the Earth, about 6,000 kilometres down, there is molten iron. This is very hot indeed – between 4,000 °C and 7,000 °C – as hot as the surface of the Sun.

The continents – America, Africa, Asia, Antarctica, Europe, Australia – move slowly around the Earth, over millions of years. Each continent floats on a plate of rock. Where these plates push against each other, they often start earthquakes. One of the worst earthquakes ever happened on 1 November 1755, in the city of Lisbon, Portugal. The earthquake continued for ten minutes, and was followed by a tsunami – a giant wall of water 15 to 30 metres high that came from the sea. Almost every house in the city was destroyed, and between 30,000 and 40,000 people died.

Another terrible earthquake hit Japan, on 11 March 2011. The earthquake took six minutes, and was followed by a tsunami which in some places was 40 metres high. It travelled far across the land, killing more than 18,000 people.

Unlike Mercury and Venus, the Earth also has a moon. Our Moon is very big; it is nearly as big as the planet Mercury. The Moon is very important for life on the Earth. As it travels around the Earth, it pulls the waters of the oceans after it. Because of this the water in all the seas of the Earth rises and falls, twice a day, following the Moon. Unlike the Earth, the Moon does not spin, so one side of the Moon is always towards the Earth.

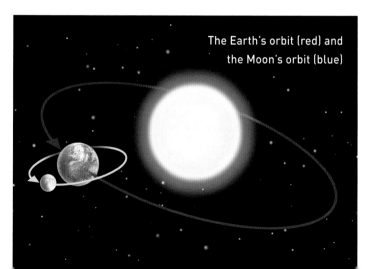

The Earth's orbit (red) and the Moon's orbit (blue)

Where did the Moon come from? How did it get there? Scientists think that once, long ago, a small planet crashed into the Earth, with an enormous BANG. The crash sent rocks and dust into space, and these slowly came together to form the Moon. In those days, the Moon was ten times closer to the Earth than it is today, and the Earth was spinning much faster. A day was only six hours long. But over millions of years, the heavy Moon made the Earth spin more slowly. Today, the Earth spins only once in twenty-four hours.

The Moon is still slowing the Earth down, and moving further away. But it is difficult to notice this. Every year, the

The small planet crashes into the Earth

Moon is 3.8 centimetres further away from the Earth. It will take 26,315 years for the Moon to move 1 kilometre further away from the Earth than it is today.

The Moon is the only place, except the Earth, where humans have ever been. But landing on the Moon is very difficult. The Moon is 380,000 kilometres away, and it is travelling around the Earth at 3,700 kilometres per hour. Much of it is covered in rocks, so it is difficult for a spacecraft to land there.

In July 1969, three men – Neil Armstrong, Buzz Aldrin, and Michael Collins – got into a small spacecraft on top of an enormous rocket, *Saturn V*. The rocket lifted their spacecraft, *Apollo 11*, into space. They travelled at enormous speed – five minutes after leaving the Earth, they were going towards the Moon at 38,000 kilometres per hour.

Four days later, they were in orbit around the Moon. Then *Apollo 11* became two different spacecraft, *Eagle* and *Columbia*. *Eagle*, with Armstrong and Aldrin inside it, began to go down towards the Moon. Collins, in *Columbia*, kept on going around the Moon, waiting for Armstrong and Aldrin to come back.

As *Eagle* came down towards the Moon, the astronauts began to have problems. The computer was not working well, and *Eagle* was coming down too quickly. Armstrong and Aldrin were 120 metres above the ground and they needed a flat place to land safely, but they could not see one. They were coming down towards some big rocks.

Armstrong turned off the computer and fired the rockets, to lift *Eagle* over the rocks. But he did not have much fuel, and there were more rocks under them. He fired the rockets again. Then, with less than a minute to go, they found a flat place.

Buzz Aldrin on the Moon, 1969

'The *Eagle* has landed,' Armstrong said on the radio. Some hours later he put on his space suit, and opened the door. Carefully, he climbed down onto the surface of the Moon. 'That's one small step for a man,' he said. 'One giant leap for mankind.'

Neil Armstrong and Buzz Aldrin were the first humans ever to walk on the Moon. At first, they walked very carefully, and their feet left clear footprints on the dusty surface. Because there is no wind on the Moon, their footprints are as clear today as they were on that day, 20 July 1969.

But the gravity on the Moon is much less than on the Earth, so the astronauts could take big steps, and jump much higher than on the Earth. Aldrin and Armstrong stayed on the Moon's surface for two and a half hours. They took lots of pictures, and picked up 21 kilograms of rocks and dust for the scientists on the Earth. Then they got back into *Eagle*. After twenty-one hours on the surface of the Moon, *Eagle* flew back to *Columbia*, where Michael Collins was waiting. Three days later, *Apollo 11* was back on the Earth.

Between 1969 and 1972, five more *Apollo* spacecraft landed on the Moon. The men on the last three – *Apollo 15, 16,* and *17* – drove around the Moon in a small car called a lunar rover. They brought back more Moon rocks and dust for the scientists on the Earth to look at. These rocks tell us that the Moon is as old as the Earth – 4,500 million years old. There used to be volcanoes on the Moon, but these stopped erupting a long time ago. Now it is just dust and rocks.

There is no life on the Moon, no air, and no water. Or at least, that is what scientists used to think. But in the last few years, they have changed their minds. Perhaps there is water after all!

There are no seas or rivers, of course. But there may be frozen water – ice. There are deep craters on the Moon where the light of the Sun never reaches the bottom. It is possible that some water has stayed there in these very cold places, frozen into the rocks.

If there is water on the Moon, that is very important, for two reasons. Firstly, water means that life is possible for humans. They can melt the ice for water to drink, and use it to grow plants. And secondly, from the water they can make

oxygen to breathe, and hydrogen to make fuel. They can use this fuel to drive cars and fire rockets. So perhaps, one day soon, scientists will go back to the Moon to live and work.

It is quite possible, today, to build a space station on the Moon. Astronauts from many different countries have been living on the International Space Station, in orbit around the Earth, since November 2000. And because the Moon's gravity is not as strong as the Earth's, it is perhaps easier to travel from the Moon to other planets, like Mars.

The *Apollo 17* lunar rover

8   Mars

For a long time, people thought there was life on Mars. In 1877, an Italian astronomer, Giovanni Schiaparelli, thought he saw some straight lines on Mars. He did not know what they were, but an American astronomer, Percival Lowell, saw them too, and decided that they were canals. There must be people on Mars, Lowell thought, and these Martians have built the canals to bring water from the icy Arctic in the north of Mars to the warmer south.

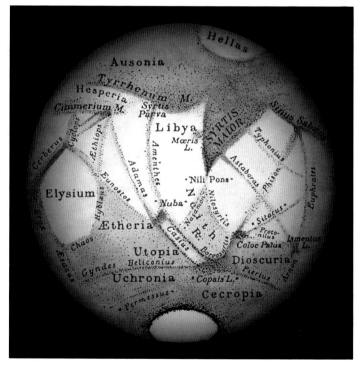

Schiaparelli's map of Mars

The idea of Martians was interesting and exciting. A book called *The War of the Worlds*, written in 1898 by H.G. Wells, told the story of Martians coming to the Earth and beginning a war. In 1938 *The War of the Worlds* was heard on the radio in the United States as a piece of radio theatre. Some people who heard it thought that it was true, and they drove away from the cities to escape the danger from the Martians.

Luckily, there are no tall grey Martians with dark eyes. When the first spacecraft flew past Mars in the 1960s, they sent back photographs of a dry place, covered with craters – as dead as our Moon. There were no Martians, no straight lines.

But then, in 1971, the spacecraft *Mariner* 9 went much closer to Mars, and found something surprising. There were no canals on Mars, but there *were* old rivers! There was no water in them now, but once, millions of years ago, they had been full of water.

So where has the water gone? And is there any there now? These are very important questions, because if there is water on Mars, then perhaps there is life there too. And if there is water, perhaps one day humans can go there to live.

Photos taken on the surface of Mars show that there *was* water there once – lots of it. The small rocks on the surface of Mars look like small rocks at the bottom of a river on the Earth. Scientists think a lot of the water floated away through the thin atmosphere into space. But they hope that some of it is still there, frozen into the rocks under the ground. If that is true, then there may already be very small life forms – bacteria – living in the rocks.

Nobody has been to Mars yet, but several spacecraft have landed there. In January 2004 two Mars rovers, *Spirit*

A Mars rover at work

and *Opportunity*, landed on Mars. These rovers are small cars full of cameras and other things that collect useful information for scientists. *Spirit* drove around for six years, travelling nearly 8 kilometres, before it stopped in soft sand. *Opportunity* was still working in 2012.

The rovers and the other spacecraft have discovered lots of interesting things about Mars. Like the Earth and Venus, Mars has an atmosphere – but the atmosphere on Mars is very different from ours. The Earth's atmosphere is 21 per cent oxygen, 78 per cent nitrogen, and just 0.035 per cent carbon dioxide. But the atmosphere on Mars is 95 per cent carbon dioxide. So if an astronaut ever lands on Mars, they will need to wear a space suit, because they need to breathe oxygen.

Because the atmosphere on Mars is so thin, the Sun only heats up the air for a few centimetres above the surface. Above that, it is quite cold. So the temperature at our astronaut's feet will be perhaps 21 °C, while the temperature on the top of their head is 0 °C. It will be summer for their feet, but winter for their head!

There are often strong winds on Mars. They sometimes blow at hundreds of kilometres an hour, making huge dust storms which cover the planet. Because the dust rises high into the atmosphere, the sky above Mars is not blue, as it is above the Earth, but red. The red colour comes from the red rock on the planet's surface. Mars is sometimes called the 'Red Planet' because of this.

Gravity is much less on Mars, because it is so much smaller than the Earth. So our astronauts will find that they can jump three times higher than they can on the Earth. That sounds like fun. But they may need that help, because they will have some high mountains to climb. The volcano Olympus Mons is 27 kilometres high, three times higher than Mount Everest, and it is the highest volcano in the solar system. Mars has some deep valleys too. Valles Marineris is

Olympus Mons

7 kilometres deep, and 4,000 kilometres long – as long as the United States.

One day soon, people will try to travel to Mars. But it will not be easy. The astronauts will have to live together in a small spacecraft for nearly a year. Then, when they land, they will not be able to walk outside without space suits. With the terrible sandstorms and icy cold temperatures, it will be a bit like living on Antarctica, but worse: on Mars there are storms of red sand instead of snow, and there is no oxygen for humans to breathe.

The first astronauts will not be able to stay on Mars for long. They will have to travel home, back to the Earth. And if anything goes wrong on the journey – if a computer stops working, or a machine breaks – no one will come from the Earth to help them. They will be on their own in space, millions of kilometres from home. If there is a serious accident, they will probably die.

But perhaps, a hundred years from now, people will build a space station on Mars. The astronauts will live inside the space station, where they can breathe oxygen without space suits. They will need to find water to drink, and to grow plants for food. They will have to put on space suits to go outside, and they will probably drive around in Mars rovers, studying the rocks of the red planet, and learning more about it. It will be a difficult, dangerous way to live.

And perhaps, after another hundred years, the first baby will be born on Mars. Human children will grow up, thinking of Mars as their home. And after a few thousand years, perhaps people will learn how to change the atmosphere of Mars, and then humans will be able to live and breathe outside.

So then there really will be life on Mars – humans like us!

Life on Mars?

9 Giants of gas and ice

On 7 January 1610, the Italian astronomer Galileo Galilei looked up into the night sky with his telescope. There was no Moon that night, so the brightest thing in the night sky was the planet Jupiter. You can easily see it on a clear night, without a telescope.

But when Galileo looked through his telescope, he had a big surprise. In the sky beside Jupiter, he saw four white spots of light. These were moons, going around Jupiter, in the same way that the Earth and the other planets go around the Sun. 'Jupiter is like the Sun,' Galileo thought. 'It has its own small planets too.'

Galileo and his telescope

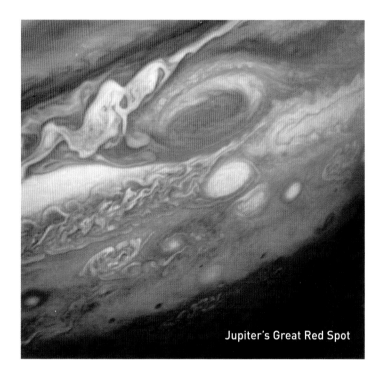

Jupiter's Great Red Spot

Galileo was right. Jupiter is almost like a small star. Just like the Sun, it is made of gas – mostly hydrogen and helium. But it has a lot more than four moons; so far, scientists have found sixty-four!

Jupiter is so bright because it is so big. It is the biggest planet in the solar system: 1,320 Earths can fit inside Jupiter. But it is much, much further away from the Earth than Mars.

On this planet there are enormous winds and storms. The Great Red Spot on Jupiter is a terrible, enormous storm bigger than the Earth. Scientists first saw this storm almost 350 years ago, and it has not stopped yet!

In 1995 Galileo went into orbit around Jupiter. But this time it was not Galileo the man, it was a space probe called *Galileo*.

As *Galileo* was travelling towards Jupiter, something exciting happened. Three astronomers, Gene and Carolyn Shoemaker and David Levy, were watching a comet. As the comet came closer to Jupiter, they saw it break into small pieces. The astronomers realized that something amazing was going to happen. 'It's going to crash into Jupiter!' they said.

And on 16 July 1994, that is what started to happen. There was an enormous explosion, and then another and another, as the pieces of the broken comet crashed into the planet. Millions of tonnes of gas exploded into space, while the *Galileo* spacecraft and the Hubble Space Telescope took photographs. Then, slowly, the gas fell back onto Jupiter again. No one had ever photographed anything like this before.

In 1995, the *Galileo* spacecraft dropped a small probe into Jupiter's atmosphere. The probe fell faster and faster, reaching the fastest speed of any spacecraft, 170,000 kilometres per hour. As it fell, it took photographs. But after a few minutes it grew hotter and hotter, broke into pieces, and disappeared.

Eight years later, there was another small explosion. The spacecraft *Galileo* followed the probe, and crashed into the planet Jupiter.

Saturn, the next planet after Jupiter, also has many surprises. Like Jupiter, it has huge storms in its atmosphere, and winds that move at 1,800 kilometres an hour. Like Jupiter, Saturn is much bigger than the Earth, but it is not very heavy.

The most surprising thing about Saturn is its rings. Galileo noticed them first, in 1610, when he looked at Saturn through his telescope. But he did not know what they were; he called them 'ears.' Then in 1655 the Dutch astronomer Christiaan Huygens looked at Saturn through a better telescope, and he said the 'ears' were really a ring. In 1675 Gian Domenico Cassini, an Italian astronomer, said that Saturn had not one ring, but two.

We know much more about these rings today, because a spacecraft called *Cassini-Huygens* has been in orbit around Saturn since 2004. The *Cassini-Huygens* spacecraft has taken thousands of photographs of Saturn and its rings. But what are these rings? Why are they there?

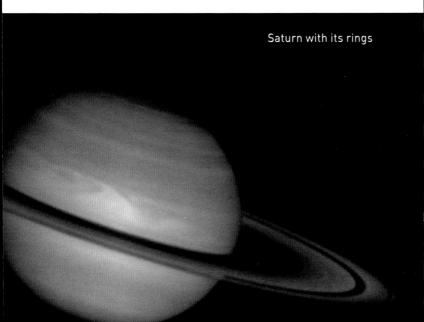

Saturn with its rings

Saturn's rings are made mostly of ice. They are hundreds of thousands of kilometres wide, but very thin – between 10 metres and 1 kilometre. There are seven rings. Some scientists think that one of Saturn's moons broke up into pieces, and the ice is what is left. Other scientists think that the rings are like the ice and dust that was there when the planets were first formed. Nobody is quite sure. But they are very beautiful.

For hundreds of years, people thought that there were no more planets beyond Saturn. But in 1781 the German-born astronomer William Herschel discovered a new planet, Uranus. This was the first planet that was discovered with the use of a telescope. And in 1846 Neptune, the eighth planet, was discovered by another German, Johann Galle, using information from the French astronomer Urbain Le Verrier, and an Englishman, John Couch Adams.

It is not surprising that it took so long to discover Uranus and Neptune, because they are an enormous distance away. Uranus is twice as far away as Saturn – about 2,869,600,000 kilometres from the Sun – and Neptune is almost twice as far again – 4,496,600,000 kilometres from the Sun.

These are enormous distances, and for a long time scientists thought it was impossible to send a spacecraft to look at them. But in 1964 a young scientist called Gary Flandro realized that for a short time around 1980, all the planets would be in a line, on the same side of the Sun. This only happens every 180 years. And because of this, Flandro said, perhaps we could send a spacecraft beyond Jupiter and Saturn, to look at Uranus and Neptune.

Two spacecraft – *Voyager 1* and *Voyager 2* – left the Earth in 1977. Two years later, they reached Jupiter. They took pictures of Jupiter and its moons, and then carried on

The launch of *Voyager 1*

towards Saturn. *Voyager 1* reached Saturn first, in November 1980, and *Voyager 2* got there in August 1981. After visiting Saturn, *Voyager 1* turned away, into deep space, but *Voyager 2* carried on, towards Uranus and Neptune.

Voyager 2 was travelling at enormous speed – about 55,000 kilometres per hour – but even at this speed it took four and a half years to reach Uranus, and another three and a half years after that to reach Neptune, on 25 August 1989. But the journey was a great success. After travelling at enormous speed for twelve years, the spacecraft arrived just six minutes later than the scientists had planned!

Sadly, Uranus is not very interesting to look at. *Voyager's* pictures showed a big blue-green planet, about fifteen times larger than the Earth. Scientists were excited, but they were not happy. They could not see through the blue clouds. Is there water down there? They do not know. They think the planet is made of frozen gas and ice, but the pictures show no mountains, no volcanoes, no storms. Just a big blue ball!

But there is one very strange thing about Uranus. Uranus does not spin in the same way as the Earth and all the other planets. Instead, Uranus moves around on its side, like a ball. And its twenty-seven moons move strangely too. Instead of going around behind the planet and coming out the other side, they go around it from top to bottom.

No one knows why this happens. Some scientists say, 'Perhaps, millions of years ago, another planet crashed into Uranus, and knocked it onto its side.' 'Well, yes,' say other scientists. 'But what happened to the other planet? Where did that planet go?' Nobody knows.

Voyager 2's photos of Neptune took four hours and six minutes to reach the Earth, travelling at the speed of light. They were more interesting than those of Uranus.

They showed a big, very blue planet with lots of clouds, like Jupiter. There was a large dark spot in the clouds – a storm, bigger than the Earth. There are winds on Neptune faster than those on Saturn, and the clouds in its atmosphere change every few minutes. But under the clouds and storms, Neptune is probably made of frozen gas and ice too, like Uranus.

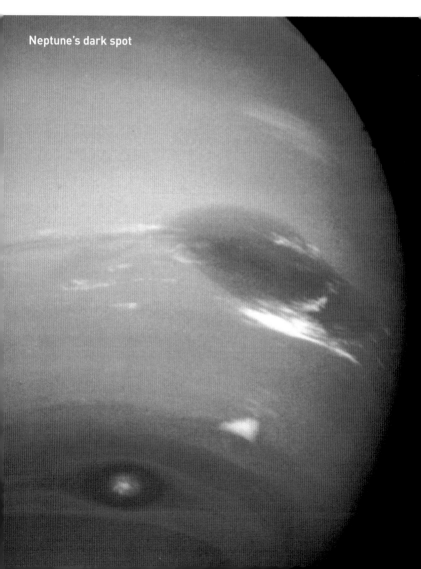

Neptune's dark spot

10 Comets and meteors

Is Neptune at the end of the solar system? No. On 18 February 1930 a young American astronomer, Clive Tombaugh, discovered another small planet, 1,400 million kilometres beyond Neptune. He called it 'Planet X', and now we call it Pluto.

But in 2006 scientists decided that Pluto is not a planet at all. Why not? It is too small, they say – much smaller than our Moon. In fact, Pluto has a moon half as big as itself. And beyond Pluto there are other rocks too, and comets.

Comets are balls of dust and ice which come from far away in space. Some come from a part of space called the Kuiper Belt, which is a very, very long way from the Earth – 6,000 million kilometres away. When *Voyager 1* was travelling through the Kuiper Belt, it took a photograph of the Earth millions of miles behind it. This famous photograph is called the Pale Blue Dot. In this photograph the Earth, our home, is so small that it is almost impossible to see. That is what the Earth looks like from the edge of the solar system.

Comets are smaller than planets, and they move differently too. Some comets spend a lot of time in the Kuiper Belt, but from time to time, they fall in towards the Sun. When they come near the Earth, we can see them in the night sky. Often they have long white tails. This happens when the sun's light meets the dust and gas coming off the comet. After they pass near the Sun, the comets fly away. Some come back a

The Pale Blue Dot

few years later, but others do not return for hundreds or even thousands of years.

Some comets are small, but others are very big indeed. In 1986, photos from the European spacecraft *Giotto* showed that Halley's comet is 16 kilometres across – a huge ball of dust and ice bigger than most cities.

Most comets go past the Earth, around the Sun, and then fly back home to the Kuiper Belt. But not always. Sometimes they get too near a planet, and crash into it, like the Shoemaker-Levy comet which hit Jupiter in 1994. When this happened, huge balls of fire were seen above the surface of the planet.

On rocky planets comets leave big craters. This has happened to all planets, billions of times. Many of the craters on the Moon and Mars were made by comets. In the night sky from time to time you may see a 'shooting star'. This line of light in the sky is made by a meteor — a piece of dust or rock, often from the tail of a comet, that enters the

Earth's atmosphere at high speed and burns up. A meteorite is a meteor that survives long enough to reach the Earth's surface. Meteorites can leave big craters on the Earth, like the giant Meteor Crater in Arizona, which is over one kilometre wide and 180 metres deep. Scientists think a huge meteorite crashed into the Earth there, 50,000 years ago.

Sometimes good things come from these big crashes. Most comets are made of ice, and some scientists think that that is why there is so much water on the Earth. Over billions of years, they say, hundreds of millions of icy comets crashed into the Earth. When the ice in these comets melted, it made some of the water in the Earth's rivers and seas.

Some scientists think that comets did another thing too. It is possible, they say, that there is life inside the icy comets. Not animals or people, of course – but very small, living bacteria. If that is true, they say, then perhaps life came to the Earth from a comet. A comet crashed into the Earth, its ice melted, and the bacteria began to grow in this new home.

Not all scientists agree with this, but it is an interesting idea. In the future, astronomers hope to send spacecraft to land on some comets, to look at their ice and rocks. Then we will know more.

11 Distant moons

But if there is life on comets, where did it come from? Everybody wants to know the answer to the question: is there life in space, outside the Earth?

Scientists do not think that anything could live on the gas giants, Jupiter and Saturn, or the ice giants, Uranus and Neptune. But all the big planets have moons, and some of these moons are very interesting. Two moons, Europa and Titan, are the most interesting of all.

Europa is one of the four biggest moons of Jupiter. Galileo was the first person to see it, in 1610. If you have a small telescope, you can see it too. It is nearly as big as the Earth's Moon.

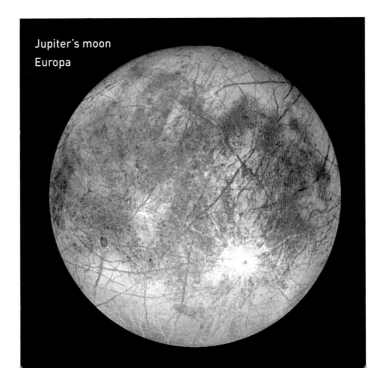

Jupiter's moon
Europa

When *Voyagers 1* and 2 flew past Europa in 1979, they saw that it was very different from Jupiter's other moons. The surface of Europa is not rocky; it is orange and white, with lots of lines running across it. In 1995 the *Galileo* space probe took a lot more photos. These photos show that Europa is covered with ice. The lines on the surface are probably places where the ice has broken. It looks very like the ice in the Arctic Ocean, on the Earth.

So what is underneath the ice? It is very cold on Europa – between -160 and -220 °C – so the ice may be 10 or even 20 kilometres thick. But underneath this ice, there is probably lots and lots of water – a huge ocean, as big as all the seas on the Earth!

To have life, you need water, but you also need heat. The heat of the Sun cannot reach the ocean on Europa, because

The icy surface of Europa

the ice is too thick. But there is another way to bring heat to the water – volcanoes. This happens on the Earth. Scientists have found volcanoes deep under the Atlantic Ocean, where there is no sunlight at all. But amazingly, small fish and animals live down there in the dark, near the volcano. So it is possible that there are volcanoes under the ocean of Europa too.

Scientists are very excited about this. One day, they hope to send a space probe to Europa. When the probe lands, it will use radar to look for water under the ice. 'We have already done this sort of thing on the Earth,' says astronomer Chris Chyba. 'There's a lake in Antarctica that's under 4 kilometres of ice. It's called Lake Vostok – and it was discovered with radar. So that's what we could do on Europa, and I think we probably will.'

If they find the ocean, they will send a submarine down into it. And that will be very exciting indeed. When the submarine turns on its lights, what will it see? Huge fish, with no eyes and enormous teeth? Nobody knows. But scientists want to find out.

Another moon which may have life is Titan. Titan is Saturn's biggest moon; it is bigger than the planet Mercury. In fact, Titan probably *was* a planet once, before it came too near Saturn. It is different from all Saturn's other moons, because it goes around Saturn the opposite way. It is the only moon with an atmosphere too. This atmosphere is mostly nitrogen, like the atmosphere on the Earth.

Titan's gravity is very low and its atmosphere is very thick. In fact, humans with wings on their arms could fly through the air there like birds. But Titan is not really a nice place for humans. The thick orange atmosphere makes it very dark, because almost no sunlight reaches the surface.

Saturn's moon Titan

So people on Titan probably could not see the Sun. It is very cold too; it is -179 °C in places.

On 14 January 2005, the *Huygens* space probe landed on Titan. It sent back more than 300 photos of hills, rivers, lakes, and seas. But the seas are not made of water: they are made of methane. There is rain on Titan too, but the methane raindrops fall much more slowly than on the Earth, and are twice as big. Titan is made mostly of water, ice, and rock, and scientists think that there may be large lakes of water or methane under the ground.

Huygens did not find life on Titan, but it is possible that something may be alive there, under the ground or in the lakes and seas. It is not a safe place for humans today. But six billion years from now, when the Sun gets bigger towards the end of its life, the temperature on Titan may rise. Perhaps then, people will go to visit it.

12 Other life in space

What about the rest of space, far beyond the solar system? Is there life out there too?

Are we alone in the universe? Or are there strange animals on other planets far away, looking at the stars and asking questions, like us? If they are there, how can we know?

NASA's *Kepler* telescope is trying to answer this question. In 2009 a spacecraft carried this large telescope into space, where it began watching more than 100,000 stars. If a planet moves in front of a star, the light from the star changes. The *Kepler* telescope sees this, and measures how big the planet is, and how far away it is from the star.

This works very well. By the end of 2011, the *Kepler* telescope had discovered more than 2,300 new planets near distant stars. Ten of these new planets were about the same distance from their star as the Earth is from the Sun. And scientists think that one of them, Kepler 22b, has water on it.

So it is quite possible that there is life on planet Kepler 22b. But what kind of life? Can we send astronauts there, to have a look?

It is a nice idea, but distances in space are enormous. Light travels at 300,000 kilometres a second, and the distance from the Earth to Kepler 22b is 600 light years – 5,676,480 billion kilometres! *Voyager 1*, the fastest spacecraft ever made, travels at 62,400 kilometres per hour. So if we send a spacecraft like *Voyager* to Kepler 22b, it will take 10,385,615 years to get there!

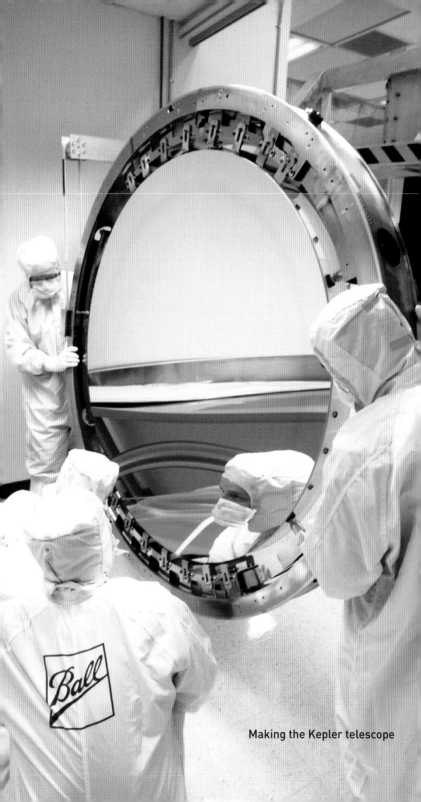

Making the Kepler telescope

Since there are so many hundreds of billions of stars in the universe, most scientists think that there must be life on many other planets – not just the Earth. But if that is true, what is living on those planets – bacteria? intelligent life like us? or strange new animals?

And if there are other planets with intelligent life, how could we talk to them, over the enormous distances of space? We can send a message at the speed of light. Kepler 22b is 600 light years away, so that means 600 years for our message to arrive, and another 600 years for an answer to get back to the Earth.

But we have sent our first messages. *Voyager 1* and *Voyager 2* left the Earth in 1977. They are still travelling, far out into deep space. They will never stop. In another 296,000 years, *Voyager 2* will pass close to another star, Sirius.

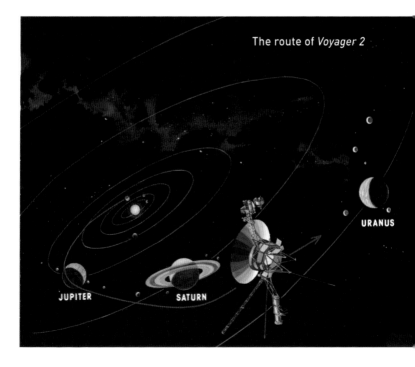

The route of *Voyager 2*

URANUS

JUPITER   SATURN

And each *Voyager* carries a golden record. On the record there are sounds and pictures from the Earth. There are also twenty-seven pieces of music, and one of them is the 1958 song *Johnny B. Goode* by rock and roll singer Chuck Berry.

In 1978 the actor Steve Martin appeared on the popular TV programme *Saturday Night Live* in the USA. He had important news, he said. *Voyager* had arrived at a distant planet, and a message had come back to the Earth – a message of four words. What could it be?

'The four words that came to us from outer space,' said Steve Martin, 'are: Send More Chuck Berry.' Across the USA, people laughed. But here on Earth, we will never stop wondering about other worlds out there. And maybe one day we really will get a message from outer space!

The golden record

GLOSSARY

amazing   surprising and hard to believe

astronaut   a person who works and travels in space

astronomer   a person who studies the Sun, planets, etc.

atmosphere   the mixture of gases around a planet

bacteria   very small things that live in air, water, earth, plants, and animals

breathe   to take air in through your nose or mouth

canal   a long narrow passage that carries water

carbon dioxide   a gas (CO_2) that people breathe out

collapse   to fall down suddenly and break into pieces

comet   a ball of ice and dust that moves around the Sun

crater   a large hole in the ground

distance   how far it is from one place to another; (*adj*) **distant** a long way away

dust   dry dirt that is like powder; (*adj*) **dusty**

earthquake   a sudden strong shaking of the ground

electricity   power that can make heat and light

enormous   very big

erupt   when a mountain erupts, it throws out burning rocks and smoke

explosion   the sudden bursting and loud noise of something such as a bomb exploding

fire (*v*)   to make an engine work

flame   a hot bright pointed piece of fire

float   to move slowly in the air

form   to grow or take shape

frozen   as cold and hard as ice

fuel   something that you burn to make heat or power

further   at a greater distance

galaxy   a very large group of stars and planets

gas   something like air that is not solid or liquid

giant (*n & adj*)   very large

gravity   the force that pulls things towards each other

helium   a very light gas (He)

huge   very big

human   a person, not an animal or machine

hydrogen   a light gas (H) that mixes with oxygen to make water

iron   a strong hard metal (Fe)

melt   to become liquid after becoming warmer; (*adj*) **molten**

methane   a gas (CH_4) that burns easily

middle-sized   of a size somewhere between big and small

nebula (*plural* **nebulae**)   a cloud of dust and gas

nitrogen   a gas (N) that is common in the Earth's atmosphere

nuclear   using energy made by splitting atoms

ocean   a very big sea

orbit   the path of a planet (or moon or spacecraft) that is moving around another thing in space

oxygen   a gas (O) in the air that people need to live

planet   the Earth, Mars, and Venus are all planets; (*adj*) **planetary**

pressure   the force with which something presses on something else

probe   a spacecraft without people on board that gets information and sends it back to the Earth

protect   to keep somebody or something safe

radar   a system that uses radio waves to find out where things are

ray   a line of light or heat

rise   to move up

rocket   a vehicle that is used for travelling into space

sand   very small pieces of rock that you find on beaches

satellite   a device in space that moves around the Earth and sends back information

scientist   a person who studies natural things; (*adj*) **scientific**

size   how big something is

spacecraft   a vehicle that travels in space

speed   how fast something goes

spin   to turn around quickly

spiral   a long shape that goes around and around

spot a small round mark

submarine a ship that can travel underwater

sunspot a dark area on the Sun's surface

supernova a star that explodes

surface the outside or top layer of something

telescope a piece of equipment that you can use to see distant things

temperature how hot or cold something is

tiny very small

total complete

universe the Earth and all the stars, planets, and everything in space

volcano a mountain with a hole in the top where fire and gas sometimes come out

war fighting between armies of different countries

ACTIVITIES

Before Reading

1 **What do you know about space? Circle *a*, *b*, or *c*.**

1 Stars are made of helium and _____.
  a) oxygen   b) nitrogen   c) hydrogen
2 The Sun is _____ million kilometres away.
  a) 12   b) 149   c) 1,692
3 The Earth, Mercury, Venus, and Mars are made of _____ .
  a) rocks   b) gas   c) ice
4 The first humans walked on the Moon in _____.
  a) 1973   b) 1969   c) 1962
5 People thought that the straight lines on Mars were _____.
  a) roads   b) rivers   c) canals
6 Saturn's rings are made of _____.
  a) ice   b) water   c) dust

2 **Three of these sentences are correct. Which ones are they?**

1 Space is getting smaller all the time.
2 Light travels at about 300,000 kilometres every second.
3 The Moon is slowly moving closer to the Earth.
4 Jupiter has more than sixty moons.
5 Everything on the Earth was once part of an exploding star.
6 Michael Collins was one of the first men to walk on the Moon.
7 The temperature at the centre of the Sun is about 6,000 °C.

ACTIVITIES

While Reading

Read Chapters 1 and 2, then fill in the gaps with these words.

astronaut, distance, galaxy, Moon, spiral, universe

1 The Earthrise photo was taken by an American _____.
2 It shows the Earth rising behind the _____.
3 A _____ is a very large group of stars.
4 The Big Bang is an idea about how the _____ began.
5 After the Big Bang, gases moved around in a _____ shape.
6 A light year is the _____ that light travels in a year.

Read Chapters 3 and 4, then circle the correct words.

1 Blue stars are *cooler / hotter* than red stars.
2 Nebulae are *clouds / explosions* of gas.
3 When a star becomes a supernova, it collapses *before / after* it explodes.
4 Betelgeuse *may / cannot* already be a supernova.
5 There are *no / a lot of* stars closer to the Earth than the Sun.
6 The corona is hotter than the *surface / middle* of the Sun.
7 It takes a little more than eight minutes for *heat / light* to travel from the Sun to the Earth.
8 We are usually protected from storms on the Sun by *sunspots / the Earth's atmosphere*.
9 When there are no sunspots, the Earth's temperature goes *up / down*.

Read Chapters 5 and 6. Are these sentences true (T) or false (F)? Rewrite the false ones with the correct information.

1 The planets were formed in the same way as the Sun.
2 Mercury, Venus, the Earth, and Mars are all made of water.
3 Mars moves around the Sun more slowly than Jupiter does.
4 A year on Mercury is shorter than a year on the Earth.
5 There are a lot of lakes on the surface of Mercury.
6 It is too hot on Venus for humans to live there.
7 The oxygen in Venus's atmosphere helps to keep the planet warm.

Read Chapter 7, then match these halves of sentences.

1 Life is possible on the Earth . . .
2 When you go below the surface of the Earth . . .
3 The earthquake in Lisbon in 1755 . . .
4 We always see the same side of the Moon . . .
5 It is hard to land on the Moon . . .
6 One of the *Apollo 11* astronauts stayed in *Columbia* . . .
7 You can still see footprints on the Moon . . .
8 Until recently, scientists used to think . . .

a while the other two landed on the Moon's surface.
b the temperature quickly becomes hotter.
c that there was no water on the Moon.
d because it does not spin.
e because there is no wind there.
f because there is a lot of oxygen and water.
g was one of the worst that has ever happened.
h because its surface is covered in rocks.

Read Chapter 8, then answer the questions.

1 What did Lowell think that the canals on Mars were for?
2 Why were people frightened when they heard *The War of the Worlds* on the radio?
3 What surprising thing did *Mariner 9* discover?
4 Why is Mars sometimes called the 'Red Planet'?
5 Why could an astronaut jump higher on Mars than on the Earth?
6 Which is the greater distance – to the top of Olympus Mons, or to the bottom of Valles Marineris?
7 How long will it take astronauts to travel to Mars?
8 What will need to happen before humans can live and breathe on Mars?

Read Chapter 9, then circle *a*, *b*, or *c*.

1 Galileo thought that Jupiter had its own _____ .
   a) spots   b) suns   c) planets
2 Jupiter's red spot is a _____ .
   a) storm   b) moon   c) volcano
3 The Shoemakers and David Levy watched a _____ crash into Jupiter.
   a) star   b) telescope   c) comet
4 The first person to look at Saturn's rings through a telescope was _____.
   a) Huygens   b) Cassini   c) Galileo
5 The eighth planet was discovered in _____ .
   a) 1610   b) 1781   c) 1846
6 In 1981, _____ travelled towards Uranus and Neptune.
   a) *Voyager 1*   b) *Voyager 2*   c) *Voyager 1* and *2*
7 Uranus _____ in a different way from other planets.
   a) spins   b) shines   c) stops

Read Chapter 10, then fill in the gaps with these words.

bacteria, comets, craters, ice, meteor, small, tail

1 Pluto is too _____ to be a planet, scientists say.
2 Comets are made of dust and _____ .
3 A comet's _____ appears when sunlight shines on dust
   and gas.
4 When comets crash on rocky planets, they leave _____ .
5 A shooting star in the night sky is made by a _____.
6 Some of the Earth's water possibly came from icy _____.
7 Some scientists think that _____ came to the Earth in a
   comet.

**Read Chapters 11 and 12, then rewrite these untrue sentences
with the correct information.**

1 Europa is twice as big as the Earth's Moon.
2 *Galileo* took photos of craters on the surface of Europa.
3 There may be volcanoes on the surface of Europa.
4 A future space probe will look for water under the ice
   using a radio.
5 Titan is the only one of Saturn's moons without an
   atmosphere.
6 The surface of Titan is very dark and hot.
7 The seas on Titan are made of nitrogen.
8 The *Kepler* telescope watches the stars from the Earth.
9 Each *Voyager* took twenty-seven pieces of food from the
   Earth into space.

ACTIVITIES

After Reading

1 Use the clues below to complete this crossword with words
from the book. Then find the hidden nine-letter word.

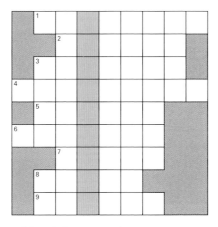

1 How far one thing is from another.
2 This moon has an orange and white surface.
3 This big planet is very blue.
4 With this you can clearly see things that are far away.
5 This is just above the surface of the Sun.
6 A huge wave that follows an earthquake.
7 This ball of dust and ice may have a long tail.
8 The name for a small car that travels around on a planet.
9 A large hole in the ground made by a rock or a comet.

The hidden word is _____.
What does this thing have to do with *you*?

2 **Who did what? Choose a name to complete the first parts of the sentences. Then match the two parts of the sentences together, and choose the best linking word.**

William Anders, Neil Armstrong, Galileo Galilei,
David Levy, Percival Lowell, Clive Tombaugh

First parts of sentences

1 _____ thought there were canals on Mars . . .
2 _____ made space history . . .
3 _____ took a photograph in space . . .
4 _____ discovered 'Planet X' in 1930 . . .
5 _____ was excited when he realized . . .
6 _____ called Saturn's rings 'ears' . . .

Second parts of sentences

7 *that / what* a comet was going to crash into Jupiter.
8 *which / who* moved water from the north to the south.
9 *although / because* he did not know what they were.
10 *as / before* the Earth rose in front of him.
11 *so / but* scientists now say it is not a planet.
12 *when / while* he became one of the first men to walk on the Moon.

3 **One word in each group (1-4) does not belong there. Find these words and move each one to the correct group.**

1 _____: corona, solar flare, solar system, volcano
2 _____: helium, hydrogen, Mercury, nitrogen
3 _____: crater, lake, sunspot, valley
4 _____: Earth, Mars, oxygen, Venus

Now choose one of these headings for each group.

gases, rocky planets, Sun, surface

4 **Do you think these things will happen in the next 100 years? Write 1-5. (1 = definitely not, 2 = probably not, 3 = perhaps, 4 = probably, 5 = definitely)**

1 We will find an ocean on Europa.
2 Astronauts will walk on the Moon again.
3 The first astronauts will land on Mars.
4 Scientists will find a planet where humans could live.
5 There will be an answer to the message on *Voyager*.

5 **Do you agree or disagree with these statements? Why?**

1 There are more important things to spend money on than space travel.
2 It is important to learn about our universe, its past, and its future.
3 We should not land spacecraft or people on other planets, because this could change things in a terrible way.
4 We should start planning now for people to leave the Earth and live on another planet at some time in the future.

6 **What is your favourite person, spacecraft, or place in space? Find some information about it and make a poster or give a short talk to your class.**

ABOUT THE AUTHOR

Tim Vicary was born in London, but he spent a lot of his childhood in Devon, in the south-west of England. He went to Cambridge University, worked as a schoolteacher, and is now a teaching fellow at the Norwegian Study Centre at the University of York. He is married, has two children and lives in the country in Yorkshire, in the north of England. Because he lives outside the city, he can see a lot of stars quite clearly in the night sky, and he has a small telescope to study the Moon and the planets.

He has written about twenty books for Oxford Bookworms, from Starter to Stage 3. His other Oxford Bookworms titles at Stage 3 are *The Brontë Story* (True Stories), *Chemical Secret* (Thriller and Adventure), *The Everest Story* and *Dinosaurs* (Factfiles), *Justice* (Thriller and Adventure), *The Mysterious Death of Charles Bravo* (True Stories), and *Skyjack!* (Thriller and Adventure). *Titanic* (Factfiles Stage 1) was the winner of the Extensive Reading Foundation's Language Learner Literature Award in the Adolescents and Adults Elementary category in 2010, and *The Everest Story* was the winner of the same prize in the Intermediate category in 2011.

Tim also writes longer books for adults. He has written three crime novels about a tough lady lawyer called Sarah Newby. These are: *A Game of Proof*, *Fatal Verdict*, and *Bold Counsel*. He has also written four historical novels: *The Blood Upon the Rose*, *Cat and Mouse*, *The Monmouth Summer*, and *Nobody's Slave*. All these books are published as e-books on the Amazon Kindle.

You can read more about Tim and his books on his website, www.timvicary.com

OXFORD BOOKWORMS LIBRARY

Classics • Crime & Mystery • Factfiles • Fantasy & Horror
Human Interest • Playscripts • Thriller & Adventure
True Stories • World Stories

The OXFORD BOOKWORMS LIBRARY provides enjoyable reading in English, with a wide range of classic and modern fiction, non-fiction, and plays. It includes original and adapted texts in seven carefully graded language stages, which take learners from beginner to advanced level. An overview is given on the next pages.

All Stage 1 titles are available as audio recordings, as well as over eighty other titles from Starter to Stage 6. All Starters and many titles at Stages 1 to 4 are specially recommended for younger learners. Every Bookworm is illustrated, and Starters and Factfiles have full-colour illustrations.

The OXFORD BOOKWORMS LIBRARY also offers extensive support. Each book contains an introduction to the story, notes about the author, a glossary, and activities. Additional resources include tests and worksheets, and answers for these and for the activities in the books. There is advice on running a class library, using audio recordings, and the many ways of using Oxford Bookworms in reading programmes. Resource materials are available on the website <www.oup.com/elt/gradedreaders>.

The *Oxford Bookworms Collection* is a series for advanced learners. It consists of volumes of short stories by well-known authors, both classic and modern. Texts are not abridged or adapted in any way, but carefully selected to be accessible to the advanced student.

You can find details and a full list of titles in the *Oxford Bookworms Library Catalogue* and *Oxford English Language Teaching Catalogues*, and on the website <www.oup.com/elt/gradedreaders>.

THE OXFORD BOOKWORMS LIBRARY
GRADING AND SAMPLE EXTRACTS

STARTER • 250 HEADWORDS

present simple – present continuous – imperative –
can/cannot, must – *going to* (future) – simple gerunds …

Her phone is ringing – but where is it?

Sally gets out of bed and looks in her bag. No phone. She looks under the bed. No phone. Then she looks behind the door. There is her phone. Sally picks up her phone and answers it. *Sally's Phone*

STAGE 1 • 400 HEADWORDS

… past simple – coordination with *and*, *but*, *or* –
subordination with *before, after, when, because, so* …

I knew him in Persia. He was a famous builder and I worked with him there. For a time I was his friend, but not for long. When he came to Paris, I came after him – I wanted to watch him. He was a very clever, very dangerous man. *The Phantom of the Opera*

STAGE 2 • 700 HEADWORDS

… present perfect – *will* (future) – *(don't) have to, must not, could* –
comparison of adjectives – simple *if* clauses – past continuous –
tag questions – *ask/tell* + infinitive …

While I was writing these words in my diary, I decided what to do. I must try to escape. I shall try to get down the wall outside. The window is high above the ground, but I have to try. I shall take some of the gold with me – if I escape, perhaps it will be helpful later. *Dracula*

STAGE 3 • 1000 HEADWORDS

*… should, may – present perfect continuous – used to – past perfect –
causative – relative clauses – indirect statements …*

Of course, it was most important that no one should see
Colin, Mary, or Dickon entering the secret garden. So Colin
gave orders to the gardeners that they must all keep away
from that part of the garden in future. *The Secret Garden*

STAGE 4 • 1400 HEADWORDS

*… past perfect continuous – passive (simple forms) –
would conditional clauses – indirect questions –
relatives with where/when – gerunds after prepositions/phrases …*

I was glad. Now Hyde could not show his face to the world
again. If he did, every honest man in London would be proud
to report him to the police. *Dr Jekyll and Mr Hyde*

STAGE 5 • 1800 HEADWORDS

*… future continuous – future perfect –
passive (modals, continuous forms) –
would have conditional clauses – modals + perfect infinitive …*

If he had spoken Estella's name, I would have hit him. I was so
angry with him, and so depressed about my future, that I could
not eat the breakfast. Instead I went straight to the old house.
Great Expectations

STAGE 6 • 2500 HEADWORDS

*… passive (infinitives, gerunds) – advanced modal meanings –
clauses of concession, condition*

When I stepped up to the piano, I was confident. It was as if I
knew that the prodigy side of me really did exist. And when I
started to play, I was so caught up in how lovely I looked that
I didn't worry how I would sound. *The Joy Luck Club*

BOOKWORMS · FACTFILES · STAGE 3

The Everest Story

TIM VICARY

It is beautiful to look at, hard to reach, and terribly difficult to climb. Winds of 200 kilometres per hour or more scream across it day and night, while the temperature falls to -20 °C or lower. Every year, some who try to climb the highest mountain in the world do not return.

But for a century people have been coming to climb Everest – some alone, some in groups, but all with a dream of going to the highest place in the world. This is their story.

BOOKWORMS · FACTFILES · STAGE 3

Future Energy

ALEX RAYNHAM

Right now, all over the world, people are using energy. As we drive our cars, work on our computers, or even cook food on a wood fire, we probably do not stop to think about where the energy comes from. But when the gas is gone and there is no more coal – what then?

Scientists are finding new answers all the time. Get ready for the children whose running feet make the energy to bring water to their village; for the power station that uses warm and cold water to make energy; for the car that saves energy by growing like a plant . . .